Other Life Exists

Wade Hobbs

PublishAmerica
Baltimore

First printing

Hardcover 978-1-4512-6133-2
Softcover 978-1-4512-6134-9
PUBLISHED BY PUBLISHAMERICA, LLLP
www.publishamerica.com
Baltimore

Printed in the United States of America

Other Life Exists

Wade Hobbs

Other Life Exists

Other life exists outside the Earth's atmosphere. Somewhere in the vast reaches of the universe, there is some form of life. The odds are literally astronomical; that much is known.

What is it to know? When a man on Earth drops a ball, he knows beforehand that it will fall. He knows this not only because of gravity, but also because of his common experience. When a woman splits a deck of playing cards, takes the middle card, places it face down, and says, "I know that this card is not the 3 of clubs," does she really know that it is not the 3 of clubs? After she turns it over and it turns out to be one of the other 51 cards in the deck, can it then be said that she really did *know* that the card wasn't the 3 of clubs?

Statistics has an answer for the last of these scenarios.

Billions of Light Years

The universe is a big place. Estimates vary, but the cosmos is generally believed to be about 13.7 billion light years old. A light year is the distance that light can travel in a year, and light travels at 186,000 miles per second. Einstein's Theory of Relativity holds that nothing travels faster than light.

In the 1940's, the late Professor George Gamow of George Washington University formulated a theory that the universe began with an initial moment of tremendous heat that is still detectable today. The theory is now known as the "Big Bang Theory," and it is widely accepted.

So even if you could build a space vessel that would move at near the speed of light, you would need to travel about 13.7 billion years to reach the other side of the universe. You probably won't live to be 13.7 billion years old. In fact, it is known that you won't live to be that old. Otherwise, you wouldn't be eligible for a pension until you were 12 billions years old.

Bacteria – Tough Enough to Live in Space

How is it known that other life lives in the universe?

There are a number of pieces to the puzzle. Bacteria in a camera left on the moon in 1967 survived until Apollo 12 returned both to Earth 2 ½ years later. During the 1980's, the National Aeronautics and Space Administration used scientific methodology to prove that bacterial spores could survive the vacuum of space. In one experiment, up to 80% of subtilis spores in the presence of glucose survived for over five years in space. These developments surprised many.

It was confirmed in 2007 that organisms are capable of surviving exposure to the vacuum of space. A microorganism called a tardigrade lived for 10 days in the vacuum. This type of creature is an extremophile, an organism that can live under radically harsh conditions that humans can't survive.

It means that if bacteria from Earth can live in the vacuum of space, the probabilities are even higher that other life is living in other parts of the universe. The probability of life elsewhere in the universe continues to climb.

Extremophiles are phenomenal. Scientists have found them living just above Lake Vostok in the ice of Antarctica. The lake is thousands of meters below the surface. The scientists have stopped drilling just above the lake.

Extremophiles prove that life is resilient. Scientists make the following argument: Extremophiles live in the harshest conditions. They survive in the vacuum of space. They live thousands of meters below the surface in Antarctica. If they can live in those conditions, they must be able to live under other conditions throughout the universe.

In light of these facts, the probability of life increases even more.

Think Like the Bug

Think of a tiny bug or microbial life living in saltwater. From the perspective of that bug, its saltwater world is radically different from the world of land. It swims around in a world made of sodium, hydrogen, and oxygen.

In any cubic foot of water taken from a sea lives microbial or microscopic life. If necessary, you can take your microscope, drive to the nearest sea, extract a cubic foot of water and analyze it. With the right techniques, you'll find microbial life.

What about other oceans, like the ocean or sea of Europa, the moon of Jupiter? Under its surface ice, that moon has an ocean that is estimated to be more than twice as big as all of Earth's oceans.

Analyze things from the perspective of microbial life. It lives in saltwater. To that tiny life, the saltwater of Europa is more similar to its environment than the dry land of Earth. When the little microbial life is swimming around in a sea on Earth, what it sees is similar to the environment that microbial life in the sea of Europa sees.

It is not entirely accurate to analyze Europa as a different world because it has some major similarities with Earth. Of course, it has some major differences, and that point is understood by most. The point that many miss is that some aspects of Europa are quite similar to aspects of

Earth, such as any given cubic foot of saltwater taken from either. Further, a cubic foot of saltwater from Earth has more in common chemically with a cubic foot of saltwater in Europa than it does with a cubic foot of air taken over dry land on Earth.

There are lethal levels of radiation from Jupiter in the vicinity of Europa. Still, it's highly probable that other forms of life are surviving in that environment. Europa may even have other forms of life unlike microbial life.

In fact, the debate over what types of life may exist in the universe has led to new departments at major universities. Astrobiology is a relatively new field dedicated to the study of life that exists in outer space. There were no independent Astrobiology Departments in the U.S. sixty years ago. Most of them are quite new.

What are the Odds?

James C. Maxwell, one of the founders of electromagnetism and a giant in physics, said, “The true logic of this world is in the calculus of probabilities.”

What are the probabilities? Basic probability theory is easy, and it arose in response to that question. When a woman pulls a card from a 52-card deck and places it face down on a table, she knows that the probability of naming that card correctly is 51 to 1.

There are certain things she can do to increase the probability of correctly identifying an individual card. For example, if she knows that new decks are stacked with reds on top and blacks on bottom, she can pull a new deck, select a card from the bottom of the deck, and predict that the card will be a particular club or spade. The odds of being correct are 1 in 26, or approximately .0385.

Statisticians, mathematicians who are usually running from their former bookies, put this in equation form in this way:

$P(R) \approx .0385$

This equation reads, “The probability of picking the right card is approximately .0385.” “R” stands for “Right card,” the card the woman predicts she has chosen. If she cuts the cards first, the probability of naming the right card is unknown because she can't necessarily divide the cards in perfectly even fashion.

If the dealer knows more, she can make a more accurate prediction. For example, if she knows further that all of the cards are stacked in numerical order within each suit beginning at the bottom, she can predict with certainty that the bottom card is either a 2 of clubs or a 2 of spades. Without further knowledge or information, that's all she can do. Her odds of naming the correct card are exactly 50/50.

$P(R)=.50$

That's all you need to know about probability theory for now.

Estimating the Odds

Will the Earth rotate tomorrow morning? Barring a major catastrophe, the odds are pretty good. So too are the odds that there is microbial life in the universe.

When you consider that the universe is very big and life can be quite small, the odds are even greater in favor of life. The odds in favor of life "go to infinity" as mathematicians say. Roughly, the odds are greater than 1,000000000000000000000000000000000 to 1. In common terms, it is known that other life exists in the universe.

Consider the waters of Europa, the moon of Jupiter. Scientists have established that an ocean or sea exists beneath the surface. That conclusion is based on at least two things. First, its surface is made of ice that has cracks in it. That suggests that water or ice has made its way through to the surface. The second fact that has convinced scientists of a saltwater sea under the surface ice is the behavior of the magnetic fields around Jupiter and Europa. The way the fields fluctuate as Europa orbits Jupiter convinces them that there is a sea inside the moon.

It was the Galileo probe that flew past Jupiter in 1995 and began collecting data in that region. The Galileo team published its findings in 2000 and concluded that an underground saltwater ocean or sea best explains the magnetic field fluctuations.

That ocean or sea is big. It is estimated to contain twice as much water as all of Earth's oceans. In the extremely far-fetched, unimaginable prospect that it contains no life, it would mean that nature has created a body of water there that is twice as big as Earth's oceans but has no life. It must contain life.

At this point, you are probably thinking, "Maybe we should revise the odds estimate to 2,00000000000000000000000000000000000000 to 1." It is almost unimaginable that a body of water that big can be insulated from microbial life. A lifeless ocean would be a first.

Instead of the big numbers just used, it's easier just to write

$$P(L) \rightarrow 1 .$$

That just means the Probability of Life P(L) goes to (→) 1, or 100%. Restated for gamblers, that means the odds are basically a sure bet:

$$O(L) \rightarrow \infty .$$

The Odds in favor of Life L go to infinity. Don't be confused by the notation. The first equation can have only a probability between 0 and 1, and the second equation has a range of 0 to infinity (∞).

Do not forget that there are three other bodies – Enceladus, Ganymede, and Callisto – that also have water under the surface. Enceladus orbits Saturn and Ganymede and Callisto orbit Jupiter. In addition, there are many asteroids that contain ice. Consider further that there are uncountable galaxies in the universe. Draw your own conclusions but don't raise form over substance.

Many tend to do that. They tend to place too much emphasis on the wrong details. It's not enough to find amino acids in space, according to people who still argue that life exists only on Earth. It's not enough to find water in so many places that astronomers lose count. Some people still argue that life is unique to Earth.

The Mystery Bacteria

Funny, isn't it, how scientific research is done!

Think about it. Scientists spend $420 billion researching the question of whether there is other life in the universe. When they find bacterial life in space, they argue about whether it was actually taken with an astronaut into space or whether it was already living in space. That may seem like a laughable scenario but it's a situation scientists face.

Some philosophers of science like George Berkeley believe that when you change your view of the universe, the universe changes. One idea connected to this concept is the problem of determining whether bacterial or extremophile life found in space was there already or whether it was carried on the space probe that reached it. In other words, when scientists find some form of microbial life in space, the issue is whether it was living there naturally or whether it arrived there with humans.

Imagine the pressure. International competition is heating up, and American scientists are feeling it. The Europeans are usually among the leaders, and it's tough to know what they've found. The Indians are in space, and the Chinese are too. The Japanese just brought a spacecraft back from an asteroid. The U.S. government has made an enormous financial commitment to science, and it wants answers.

At that point, you don't just walk into the Oval Office and say, "I'm sorry, Mr. President. I'm not sure if the bacterial life we found was actually from space or whether we took it with us."

That's why the lab leaders hire hundreds of post doctorates who, like the bacteria they study, live off meager rations such as peanut butter and jelly. Don't feel sorry for post docs—there's a minimum PB&J requirement they must meet before becoming lab leaders. Don't feel sorry for the bacteria either. The post docs often carry their bacteria with them when they're promoted. It's a team effort.

Scientists have a very serious debate. Consider the example of the weather balloon that was sent up over India. It was sent to the very edge of space. The best among scientists still debate whether the microbial life they captured was taken from the edge of space or whether it was taken from the Earth's surface up to the capture point. The scientists who conducted the experiment are convinced that they did the experiment correctly and that the captured microbial life was actually from the edge of space.

Applying the Rules

When you think about rocks in outer space, you apply the rules of biology, so why stop? It works.

When you think of life on the moon, you think of a man in a protective suit hopping around having fun. You certainly don't think of bears, rhinos and tigers. The rules of biology are at work. Go forward to the surface of Europa. You obviously don't think of cats and mice running around on the surface ice unprotected. There is a scant environment on Europa, but scientists know it's not enough to support human life. Logically, they infer that it can't support mammal life either. The rules of biology are at work.

Why would you ever want to abandon those rules?

The rules—or heuristics—of biology are at play just like the rules of the other hard sciences, physics and chemistry. You don't expect the physicist at the Jet Propulsion Lab to abandon the rules of physics when she calculates a trajectory around Jupiter. You don't expect another physicist to abandon the rules of physics when he estimates the electromagnetic field around Europa. It was physics that allowed scientists to reach Jupiter, so you don't expect them to abandon those rules.

You don't expect those scientists to abandon the rules of chemistry when they look through their spectrographs. Functionally, a spectrograph is used to discern the types of chemicals that are in a certain place by measuring the wavelength of the light that has passed through the chemical.

Astronomers use spectrographs to determine what types of chemicals are in the universe. Consider the example of distant gas in outer space. Behind the gas is a light emitting object like a star. The light from the star passes through the gas and reaches Earth, where a post doctorate is sitting with a spectrograph. The post doc puts away the sandwich and looks into the spectrograph, which is basically a device that divides the light. Each gas creates a different signature in the spectrograph.

The rules of biology work every day in this universe. Obviously, no scientist has ever discovered a tiger, alligator, eagle or rhino on any other rock in the universe except Earth. It would be vain to abandon the rules of biology and search for such mammals in an environment that doesn't support them, such as the moon.

If those biology rules allow you to immediately eliminate the possibility of finding a non-human mammal on the moon, you shouldn't abandon them when you consider the issue of whether there is other life in the solar system. Apply the familiar biology rule, "Where there is water there is life." Scientists apply this rule and conclude that there is probably life in the waters of Europa. The statistician walks up and says, "Yeah, and there are so many water sources that you know there is life somewhere in the universe."

Think of this scenario. You have a 100 to 1 bet that the next poker chip someone will hand you will be red. The odds are in your favor so of course you take the bet. Then, someone else who is in the know tells you that there are only 10,000 chips in the entire place, and only one of them is blue. You place the bet that the next chip you receive will be red because the odds are so tilted in your favor that you know the next chip will be red.

That's roughly the situation with the prospects of finding life in the universe. That's why you know other life exists in the cosmos.

Skeptics, Reason and Truth

There are skeptics out there, so it's best to start with a fair argument. Life either exists on or in Europa or it doesn't. So assign a 50% probability to each possibility. For the starting point of the debate, assign a probability p of .5 to the possibility that life exists there and a probability x of .5 to the probability that it doesn't.

Few scientists would quarrel with this initial assignment of probability.

Reason acknowledges the conclusions of the Galileo team that Europa must have an ocean or sea under its surface ice. That's the most logical explanation given the behavior of the electromagnetic fields. Build on that valid conclusion.

A good thinker would necessarily apply more probability to the argument in favor of life in Europa simply because that moon has water. To do otherwise is to reject the laws of biology. The only real issue is how much probability to apply. A statistician might apply as much as 50% more to the prospects of life, or as little as 1% might be applied. Anything less than 1% is a trifle.

Scientists can quibble about the value between 1% and 50% as much as they like. They can even quibble or argue about whether 1%, or .01 if you like, is statistically large enough to be significant. Scientists must accept, however, that some additional probability must be applied to the probability of life in Europa. They abandon reason otherwise.

Perhaps there is a scientist who might argue, "Oh, gee, I realize that there is abundant life throughout the seas of this planet and that there is plenty of water in Europa. But maybe those facts should be given no thought whatsoever." Maybe you should disregard that scientist.

The reason a good scientist would assign even more probability to the probability that life exists in Europa is the biology heuristic, "Where there is water there is life."

Again, a good thinker might begin with these equations:
p = .5 (the probability that there is life in Europa, or 50%)
x = .5 (the probability that there is no life in Europa, or 50%)

After the initial adjustment, those equations are understood in these terms:
p is greater than or equal to .5, and
p is less than or equal to 1 (or 100%)

x is greater than or equal to 0 (0%), and
x is less than .5 (50%).

It's only logical to drop x down to .49 as an initial step, but scientists are prone to quibble. Given the biology rule, "Where there is water there is life," it's rational to drop the x probability down at least 1%. Anything less gives a false view of how common it is to find life in water.

You might even reduce the x probability down to 0 based on the fact that in every natural, unpolluted body of water that has been explored, life has been found.

Scientists have yet to find a natural body of water on Earth that has no life. Based on the strength of these assertions, it is rational to reduce the probability of a lifeless Europa by somewhere between another 1% and 39%. That brings the x probability against life down to 10% and the p probability in favor of it up to 90%.

Consider further that 122 seas have been examined rather than just a few. The probability p that life exists in Europa begins to approach 100%.

Building Knowledge

What's new here, to the extent that anything can be original, is knowledge. This isn't a matter of "There is probably life in Europa." Nor is it a matter of "There is an overwhelming possibility that life exists in Europa." It is not "I feel like there is life in Europa," or "It wouldn't surprise me." Nor is it the old routine about "a strong possibility that life exists."

It is known statistically that life exists in Europa. That's the advance. It hasn't been physically confirmed yet but it obviously exists.

Consider the Earth's oceans. There are either 5 or 7 depending on how you count them. The Atlantic, the Pacific, the Indian, the Southern, and the Arctic are the 5 major oceans, and the Atlantic and Pacific are often divided as North and South. There are 122 seas if you count them correctly. Some seas are actually part of an ocean, but don't let that bother you. The number 122 works better for the argument, and you never let anyone screw up your argument.

Consider the seas in the world. There are 122 of them, including the Gulf of Mexico, the Dead Sea, the Mediterranean Sea, and 119 others. The Dead Sea is included because it has microbial life. Ignore bodies of water like the Great Lakes in the United States because they are freshwater lakes and their chemical content is therefore different.

In layman's terms, the odds are at least 122 to 1 in favor of there being life in Europa. In fact, they are better than that. Every time a body of water this big has been found, life also has been found. The only exceptions are other bodies in outer space that have not been explored yet. Those bodies include Enceladus, Ganymede, Callisto, and three planets in the Upsilon Andromedae Star System.

That life exists in Europa can be known through the laws of statistics.

Scientists don't abandon the laws of physics whenever they send a space probe to Jupiter. They apply the laws of gravity and electromagnetism just as they apply them here on Earth. They don't abandon the laws of chemistry; they apply those laws when they look through spectrographic data. Why should they suddenly abandon the third branch of the hard sciences, biology? There is the familiar heuristic in biology that reads, "Where there is water there is life."

Opponents of this view would like to say, "Abra-ca-dabra, the rules of biology do not apply." Such incantations do not work. There is no rational reason to abandon the laws of science, including the laws of physics, chemistry, and biology. Some apparently expect to find an ocean with twice as much watcr as all thc occans on Earth and conclude that it has no life. That expectation is inconsistent with the biology heuristic just mentioned.

Some tardigrades survived ten days of exposure to the vacuum of space. Image courtesy of the Astrobiology Institute, National Aeronautics and Space Administration.

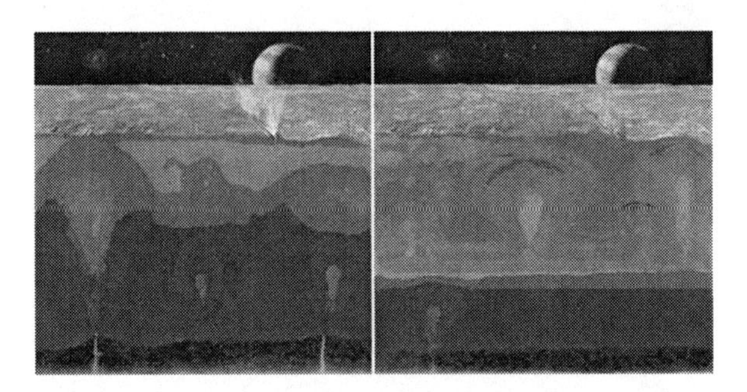

The precise dimensions of the sea beneath the surface ice of Europa are unknown. Shown are two possible scenarios. Image courtesy of the Jet Propulsion Lab and National Aeronautics and Space Administration.

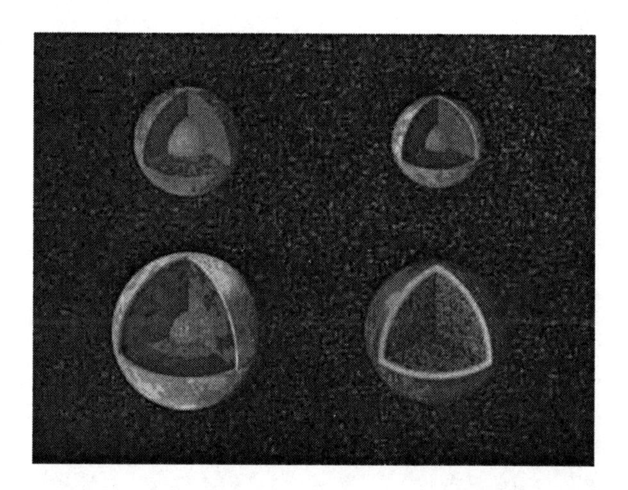

This drawing shows what scientists believe to be the internal structures of the Galilean moons of Jupiter. Io, in the upper left, is the only one believed not to have a sea beneath its surface. From top right, the other three moons are Europa, Callisto, and Ganymede. The four moons are called "Galilean" because Galileo first discovered them. Image courtesy of the Jet Propulsion Lab and the National Aeronautics and Space Administration.

Images of Europa taken by the Galileo probe. Images courtesy of the Jet Propulsion Lab and the National Aeronautics and Space Administration.

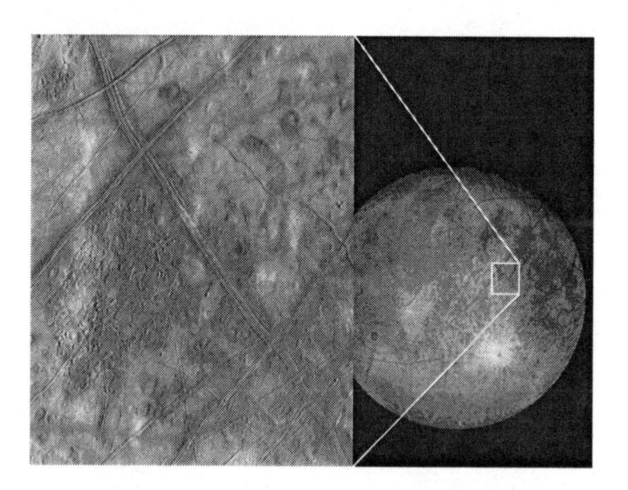

Another image of Europa. The inset shows cracks in the surface ice. Image courtesy of the Jet Propulsion Lab and the National Aeronautics and Space Administration.

Shown are the magnetic fields around Europa and Jupiter. The Outer Planet Flagship Mission is a joint project between the National Aeronautics and Space Administration and the European Space Agency to send a probe to study those fields. The fluctuation of the magnetic fields proves that Europa holds an ocean of saltwater beneath its surface. Image courtesy of the Jet Propulsion Lab and the National Aeronautics and Space Administration.

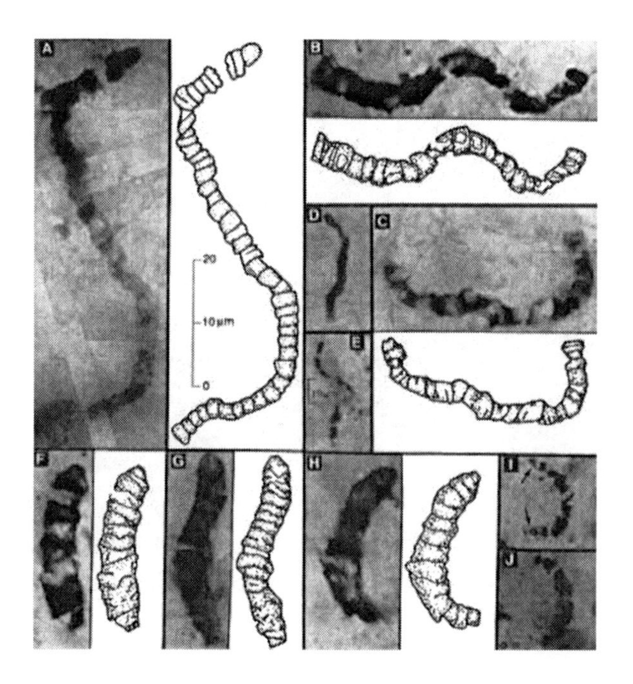

Another example of bacterial or microbial life from Earth. Image courtesy of the National Aeronautics and Space Administration.

Plankton or microbial life that lives in the oceans of Earth. Image courtesy of the National Aeronautics and Space Administration.

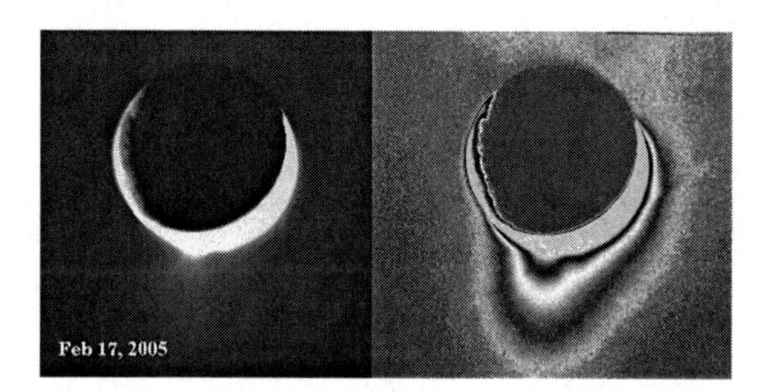

These images show the plume of icy material spewing from the South Pole of Enceladus, a moon of Saturn. Image courtesy of the Jet Propulsion Lab and the National Aeronautics and Space Administration.

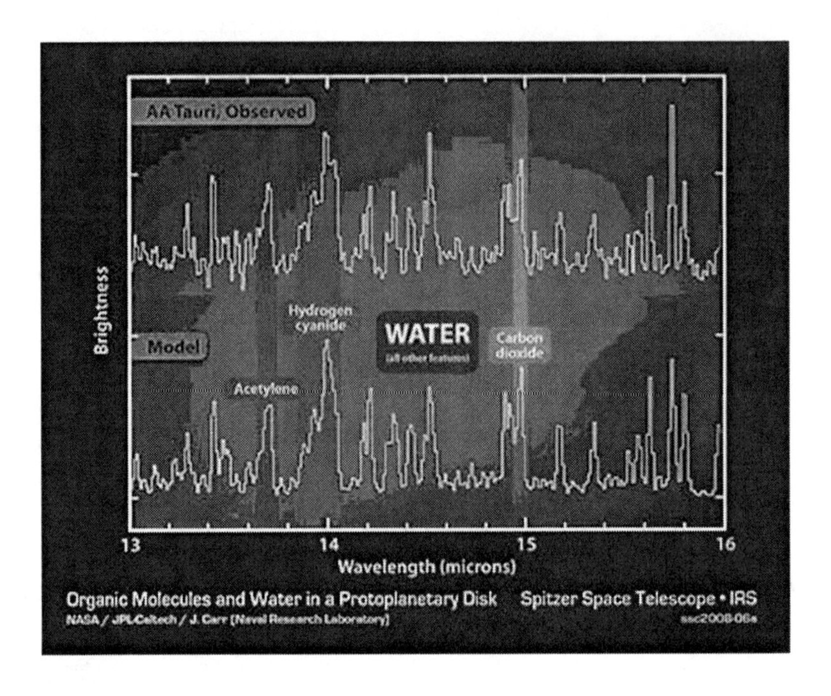

Scientists use the Spitzer Space Telescope to search for organic molecules and water. Image courtesy of the Jet Propulsion Lab, CalTech, the National Aeronautics and Space Administration

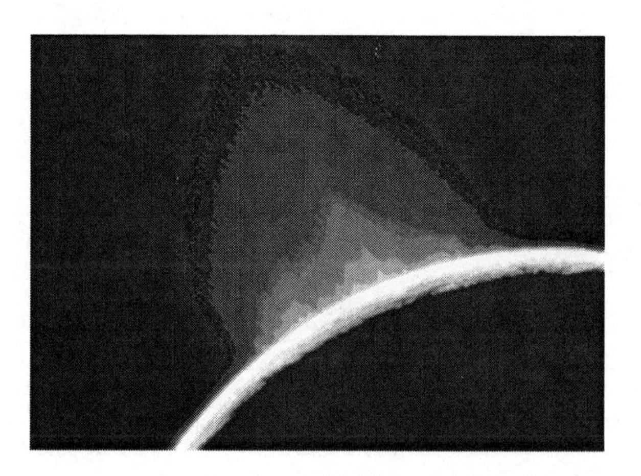

The icy plumes of Enceladus, moon of Saturn. Image courtesy of the Jet Propulsion Lab, the National Aeronautics and Space Administration, and the European Space Agency.

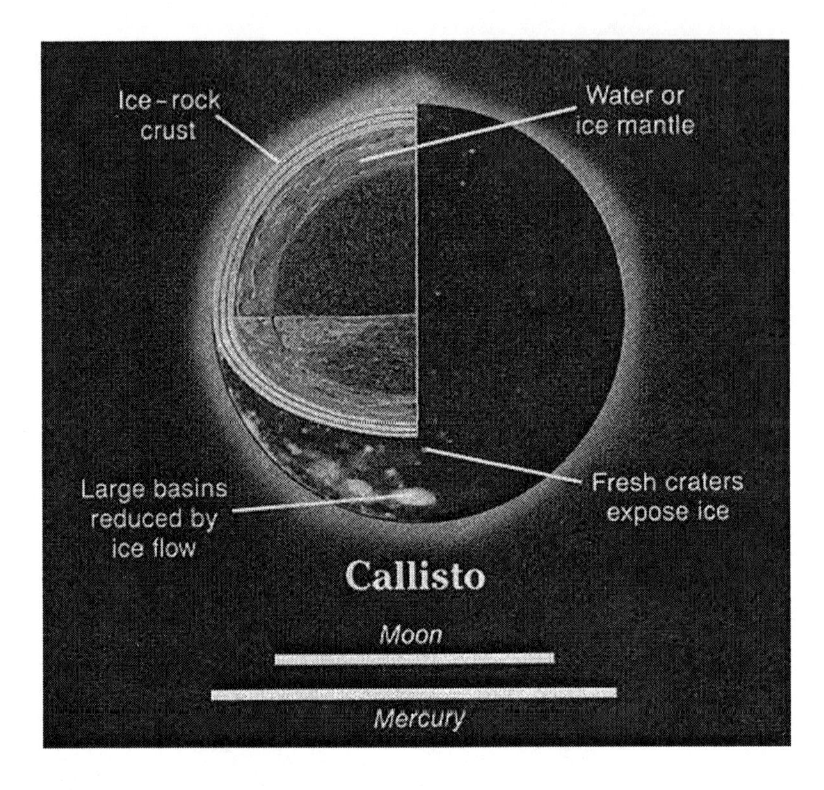

Callisto has water beneath its surface ice. The relative sizes of the Moon and Mercury are also shown. Image courtesy of the Jet Propulsion Lab and the National Aeronautics and Space Administration.

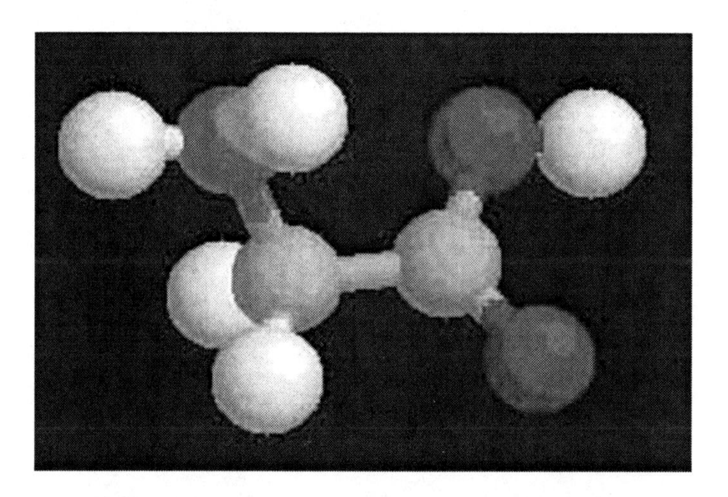

The Stardust probe found Amino acids in the tail of a comet. The probe flew through the comet's tail, collected samples, and returned to Earth in 2006. Image courtesy of the Goddard Space Flight Center and the National Aeronautics and Space Administration.

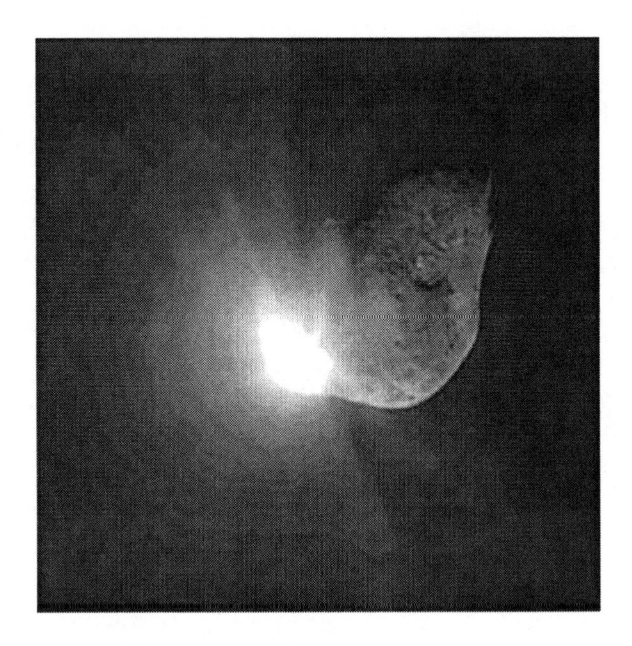

The Deep Impact Mission produced this photo of the comet Tempel1.
Image courtesy of the National Aeronautics and Space Administration.

A mosaic image of Eros. Image courtesy of the Goddard Space Flight Center and the National Aeronautics and Space Administration.

Initial Conditions

It may seem like a leap of logic in this argument to equate the initial conditions for life in saltwater on Earth and saltwater in Europa.

In fact, it's quite logical. Just consider the chemical content of two samples of saltwater, one from Earth and the other from Europa. One has sodium, hydrogen and oxygen. The other has sodium, hydrogen and oxygen. What a miraculous human achievement! You win the gold medal!

Saltwater near the surface of an Earth ocean receives energy from the sun. Saltwater in Europa receives energy from the core.

Roger, Houston. There is life in Europa.

Compare the saltwater taken from an Earth sea to the air on dry land. The saltwater consists of hydrogen, oxygen, and sodium. The air is hopefully oxygen rich. It should be oxygen rich. In any case, the chemical content is quite different.

Deductions

This isn't one of those "There's life out there, I just know it" arguments. You hear such arguments occasionally. They come from people who seem to have a religious belief or something of that nature that life is out there. When asked why they believe life exists in outer space, they seem to have little but a vague recollection of a television program. When asked about the details, they are hard pressed to recount what they've seen.

Consider the mathematical proof in the form of statistics. Logical reasoning based on biology supports that proof.

The problem for skeptics is the numbers. It is known that numerous asteroids have water ice. It is known further that numerous bodies in outer space have water or water ice. The exact number of objects bearing water or water ice in the universe is unknowable because there are so many of them.

Science has established that most asteroids are made of rock and ice. The ice is often water ice rather than frozen hydrogen or some other element. The current limitation on science is that it cannot discern the specific chemical content of asteroids outside the solar system. It is not necessary to do that!

When you sail off the Atlantic coast five miles, you don't need to dive to the bottom to know that there is sand on the ocean floor. You already know that based on your experience and the natural laws. You've gone swimming off the beach and sand covers the bottom. You've seen crabs and lobsters in restaurants that still have sand on them. You know from physics that the moon creates tides that move massive amounts of sand around the ocean floor. You infer the existence of sand on the ocean floor.

So too can you infer the existence of asteroids bearing water or water ice in other star systems.

That is true even though you can't see the asteroids or collect data on them with current telescopic equipment. Just as you don't need to physically see the bottom of the ocean floor to know there is sand there, you don't need telescopic views of distant asteroids to know that a large percentage are composed of rock and ice. You have seen pictures in books and read about their content. There have been numerous asteroid missions over the past decade, including Eros, Dawn, Rosetta and Hayabusa.

In fact, the term "asteroid" has traditionally been used to describe small rocks in the solar system. More recently, some astronomers have dubbed the term "small Solar System bodies." You can use the term "asteroid" to describe all the rocks orbiting the sun and other star systems.

The Mars Controversy
and Other Bodies that Have Water

Some controversy exists as to whether life was discovered on Mars by the Viking Mission. A minority of scientists believe that the 1976 Viking experiment found living microorganisms in the soil of Mars.

To learn more about things like LR, GC-MS, CO2, TV-GC-MS, and the building blocks of life, read about the Viking experiment.

Most sources present the conclusion that life was not found on Mars. The New York Public Library take is that Viking 1 is the first spacecraft to hunt for life beyond the Earth and moon. The Sci-Tech Department of the Carnegie Library sums up the issue of whether life was discovered on Mars: It's inconclusive.

Some believe that of the planets, Mars is the most likely to harbor life.

Aside from Europa, Ganymede, Callisto, and Enceladus are three bodies that contain water. Ganymede and Callisto are moons of Jupiter and Enceladus is a moon of Saturn. Enceladus actually has plumes of water spewing from it. According to the BBC, the Cassini Mission has confirmed the basic building blocks of life there, including water, energy and sodium, even organic molecules.

Is This the Guy for Us?

You can imagine what will happen when scientists finally reach the conclusion that there is other life in the universe, and consider the particular case of Europa. Sometime after they conclude that there is life there, a few of them may find themselves in a room with a man who has his head down on a table.

"Are you all right?" they may ask him.

They may nudge him a bit to wake him.

A sign that reads "Navy" will hang on the door outside.

"Sailor, what time did you get in last night?"

"About fifteen minutes ago" will probably be the response.

"Maybe we shouldn't have given them so much booze," the scientists will probably think to themselves.

"Sailor, we have a mission to Europa we'd like you to consider. We think there's other life in the universe."

"What do their women look like?" will likely be the sailor's drunken response.

"Are you sure this is the guy for us?" the scientists will probably say.

"Maybe we should get somebody else."

Small Amounts of Math are Painless

The mathematical statement of the argument for other life in the universe is in the little section at the end called "A Little Math for a Simple Problem." In it you'll find the equation that describes the conditional probability of finding life (L) given water (W). Generally speaking, the probability of finding life (L) given water (W) equals 1. That's the same as saying, "the probability of finding life (L) given water (W) is 100%."

That's true, all other things being equal. You may find an unnatural water source that man has polluted so much there is no life. Still, if you take 100 samples of say, a gallon each, there's an excellent probability that you'll find life.

The second equation on that page may be seem contradictory but it's really simple. It reads, "The probability of finding life (L) given water (W) is .995, which is basically 1." Of course, a probability of 1 is the same as 100%.

The second equation is a heuristic. It's included because out of 10,000 small samples taken from Earth, you might find a few that have no life. By that time, you're probably violating the "all other things being equal" provision above by introducing human manipulation into the experiment.

Easy Ways of Calculating Probabilities

The second section at the end that you'll probably want to read is called "More Beautiful Equations." The first equation is just a two-condition probability equation. That means that it states the probability of finding the first term given the second and third terms.

The first equation states the probability of finding life (L) given water (W) and sodium (S). Reduced to its simplest form it reads, "The probability of finding life (L) given water (W) and sodium (S) goes to 1."

The second equation is just a restatement of the first.

A Probability Argument

Here's the strongest probability argument that life exists in places other than Earth. Perhaps it's a bit redundant but there are so many skeptics it seems necessary.

In places where there is natural water on Earth, the probability is over 99% of finding life, at least in microbial form. Water is constantly being discovered throughout the solar system and the Milky Way galaxy. You don't discard the rules of biology just because you leave Earth.

The Oort Cloud has innumerable objects in it, and most of them are comets or asteroids that eventually fall from their orbit and approach the sun. When the asteroid or comet approaches the sun, the ice begins to melt and it sometimes becomes a comet that is visible from Earth. For a comet with only a small amount of water, it may not even be visible from Earth. There is little data on the rocks in the Oort Cloud, so even though it's possible that some do not have water, it is difficult to predict how many have water and how many do not. The best a skeptic can do is assign a 50/50 probability to whether a given Oort Cloud rock has water ice or not.

Currently there is no scientific data that definitively estimates the number of objects in the Oort Cloud. Even if it only consists of 100 million objects, which is a low estimate according to some, the odds are ridiculously in favor of life of some type on one of those rocks. Granting 50/50 odds of each rock having water, which seems generous to the skeptics, there would be an estimated 50 million rocks with water in the Oort Cloud. The skeptics must prove that none of those rocks bear life, known or unknown in form. It's at least 50 million to 1 in favor of water, and that's only in the area near the sun.

Consider the other rocks in the galaxy and the universe, and the number of places that may bear life increases exponentially from 50 million.

Whenever you get a bet that's at least 50 million to 1 in your favor, you take it.

Looking at Asteroids, Comets
and Other Objects

A spectrograph allows scientists to analyze chemical content. They are quite useful to astronomers.

There are numerous objects around the sun that have water ice. The Oort Cloud, originally hypothesized by Dutch astronomer Jan Oort in 1950, is an immense body of comets that surrounds the sun. Gravity theory confirms the Oort Cloud's existence. Oort believed that comets were attracted to the sun by gravity and that they returned to the cloud of comets after flying by the sun. Computer simulations of comets have confirmed his theory. Still more objects near the sun have water ice.

Callisto, a moon of Jupiter, has ice. Uranus and its moon Miranda have ice, as do Neptune and its moon Triton. Generally, most of the rocks out there have some form of ice because it's so cold.

Asteroid Missions

Astronomers distinguish between asteroids and comets. Rocks smaller than planets were traditionally called asteroids. Recently there has been a debate about what to call asteroids, but don't worry about it. People usually find something to argue about and there's little you can do about it. Remember – it's not your fault.

Comets are essentially just ice-rich asteroids that are spraying a trail of ice as they approach the sun and melt. They continue to melt for some time after passing the sun. The Kuiper Belt of asteroids is a source of many comets as is the Oort Cloud, a body of space rocks that stretches far into space.

Eros is an asteroid. During the Near-Earth Asteroid Rendezvous Mission, the National Aeronautics and Space Administration sent a probe to visit it. The probe orbited Eros for a year beginning in 2000. In 2001, it landed on the asteroid and collected data. Eros is basically a big rock shaped like a potato.

Dawn is another mission from the National Aeronautics and Space Administration. Its mission is to seek out and explore asteroids Vesta and Ceres; to boldly flyby what no other probe has flown by before. They may even make a television show about it.

Rosetta is a European Space Agency mission to asteroids, and Hayabusa is a Japanese mission that had its problems. With hopes of collecting dust samples, the latter had a rendezvous with an asteroid and returned to Earth in June 2010. As of this writing, the Japanese space agency had not published any claims of finding living organisms on the Hayabusa spacecraft.

Question Answered

You sometimes read things like, “Since the beginning of recorded history, man has pondered the question, ‘Is there other life in the universe?’”

Well, now you know. For centuries man pondered the question. He knows. It exists. You’ve got it.

And if you’re into laughter, you’ll probably like “Hitchhiker’s Guide to the Galaxy,” by Douglas Adams, who wrote, “You just come along with me and have a good time. The Galaxy’s a fun place. You’ll need to have this fish in your ear.”

You can either read him and have fun or you can continue to debate the ancient question of whether there is other life in the universe.

Life in the Universe

The question has been answered. Is there other life in the universe? The answer is "Yes."

To review the arguments, consider Europa again. It is known that life exists in Europa because of statistics. The standard in statistics is to establish a significance level of .05 or 5%. Statisticians establish a null hypothesis and an alternative hypothesis for mutually exclusive propositions.

For Europa, life either exists or it doesn't. Starting with a 50/50 probability for each proposition, the "no life" proposition is rejected because of valid statistical analysis. The invalid theory is rejected and the valid conclusion, that life exists there, is accepted. The mystery is solved. You can find the statistical argument in the appendix entitled, "Finding Life in Europa."

The same arguments apply to Enceladus, Ganymede, and Callisto. Enceladus orbits Saturn, and Ganymede and Callisto orbit Jupiter. Each moon has water beneath the surface.

The other argument is really easy. New sources of water are being discovered each year, if not each day. Where there is water, there is life. That's an established maxim in biology, and there's no reason to begin rejecting the rules of science. Consider the solar system alone. The Oort

Cloud, a body of asteroids and comets that surround the sun, contains so many comets they can't be numbered. Comets contain water. The odds in favor of life are so big that they are inestimable.

When you consider the billions of light years of space that contain innumerable water-bearing rocks, the odds in favor of life are "beyond inestimable." Consider further that extremophiles can live in space and there is only one conceivable conclusion.

It is known statistically that other life exists in the universe.

Appendix 1:

A Little Math for a Simple Problem

$$P(L|W) = \frac{P(L \cap W)}{P(W)} = 1 = 100\%$$

This equation translates to "Where there is water there is life." L stands for life and W stands for water. P(L / W) is the probability of life given there is water.

$$P(L|W) = \frac{P(L \cap W)}{P(W)} = .995 \approx 1$$

This is the same rule in heuristic form. There is only a minute difference between the two equations. The second equation merely recognizes the possibility that you may find some water on Earth where there is no life, but the probability of finding life in water is so high it is estimated at 99.5%, or .995.

Appendix 2:

More Beautiful Equations

$$P(L|W,S)= \frac{P(W|L,S)\, P(L|S)}{P(W|S)} \rightarrow 1.$$

This equation reads, "The probability of life (L) given the presence of water (W) and sodium (S) equals the probability of water (W) given the presence of life (L) and sodium (S) times the probability of life (L) given sodium (S), divided by the probability of water (W) given sodium (S). That probability approaches 1.

Alternatively, this can be stated

$$P(L|W,S)P(W|S)=P(W|L,S)\, P(L|S)$$

Appendix 3:

Finding Life in Europa

Calculations for Life in Europa or a Body Like Europa

$p = .5$ *(the probability that there is life in Europa, or 50%)*
$x = .5$ *(the probability that there is no life in Europa, or 50%)*

Now, you adjust those equations to reflect the biology rule, "Where there is water there is life."

$1 \geq p \geq .5$
$0 \leq x < .5$

Restated to three decimal point accuracy:

$1 \geq p > .499$
$0 \leq x \leq .499$

Be formal and establish the null hypothesis as:
H_0 is the proposition that no life exists in Europa.
Establish the alternative hypothesis as:
H_1 is the proposition that life exists in Europa.

Roughly equate the initial conditions between naturally occurring saltwater on Earth and naturally occurring saltwater in Europa.

$S_E \approx S_{EU}$

Consider all the seas on Earth that have life. There are 122 seas, and 122 of them have life. Roughly equate the sea of Europa to an Earth sea. Your probability range is between 1 and .992 in favor of life.

$122/123 = .992$ and
$.992 \geq p \;\; 1$ or
$\mathbf{p \rightarrow 1}$

The probability p of life in Europa approaches 1.

Reject the null hypothesis.
Accept the alternative hypothesis.

CPSIA information can be obtained at www.ICGtesting.com
Printed in the USA
LVOW11s0247170614

390288LV00001B/92/P

9 781451 261349